About Skill Builders Word Problems

by Clareen Arnold and Nancy Bosse

Welcome to Rainbow Bridge Publishing's Skill Builders series. Like our Summer Bridge Activities collection, the Skill Builders series is designed to make learning both fun and rewarding.

Students often ask their parents and teachers, "When am I ever going to use this?" Skill Builders Word Problems books have been developed to help students see the many uses of math in the world around them. Content for this book is based on current NCTM (National Council of Teachers of Mathematics) standards and supports what teachers are currently using in their classrooms. Word Problems can be used both at school and at home to engage students in problem solving.

The second-grade math skills used in this book include addition, subtraction, multiplication, measurement, time, and money values.

A critical thinking section includes exercises to help develop higher-order thinking skills.

Learning is more effective when approached with an element of fun and enthusiasm—just as most children approach life. That's why the Skill Builders combine entertaining and academically sound exercises with eye-catching graphics and fun themes—to make reviewing basic skills at school or home fun and effective, for both you and your budding scholars.

Table of Contents

Planning a Field Trip to the Zoo

Your class is planning a field trip to the zoo on February 12. Use the calendar to answer the questions below.

February

Sunday	Monday	Tuesday	Wednesday	Thursday	Friday	Saturday
1	2	3	4	5	6	7
8	9	10	11	12 Field trip!	13	14
15	16	17	18	19	20	21
22	23	24	25	26	27	28

1. What day of the week is February 12th?

   _____

2. Draw a heart on February 14th for Valentine's Day.

3. How many Saturdays are there before the field trip?

   _____

4. How many days must you wait before you go to the zoo if today is February 1?

   _____

5. How many days are there between the zoo trip and Valentine's Day?

   _____

3

Use the calendar to answer the questions below.

July						
Sunday	Monday	Tuesday	Wednesday	Thursday	Friday	Saturday
	1	**2**	**3**	**4**	**5**	**6** Bird Show 3:00
7	**8** Radical Reptiles 7:30	**9**	**10** Radical Reptiles 7:30	**11**	**12** Radical Reptiles 7:30	**13** Bird Show 3:00
14	**15**	**16**	**17** Dolphin Show 7:00	**18**	**19**	**20** Bird Show 3:00
21	**22**	**23** Large Animal Show 2:00	**24** Dolphin Show 7:00	**25** Large Animal Show 2:00	**26**	**27** Bird Show 3:00
28	**29**	**30**	**31** Dolphin Show 7:00			

1. What day of the week can you watch the Dolphin Show?

_____

2. If you are at the zoo on a Saturday in July, what animal show can you see?

_____

3. If you are at the zoo on July 9, will you be able to see an animal show?

_____

4. What time is the Large Animal Show on July 23?

_____

A Race in the Air

Read the clues below. Draw each bird in its correct square.
Write its ordinal number.

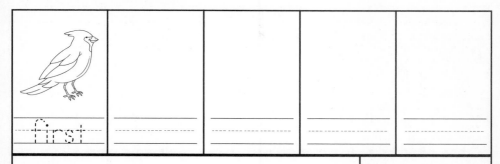

first

Picture Bank

The cardinal is first in line.

Beside the cardinal is the toucan.

The duck is fourth in line.

The hummingbird is in front of the duck.

The turkey is last in line.

Word Bank

first

second

third

fourth

fifth

Draw a picture.

Who came in second?	Who came in fourth?

Word Problems Grade 2—RB–904012

The Race in the Bear Den

Read the clues below. Draw each bear in its correct square. Write its ordinal number.

sixth				

The black bear is sixth in line.

The teddy bear is last in line.

In front of him is the polar bear.

The koala bear is behind the black bear.

The panda bear is eighth.

Picture Bank

Word Bank

~~sixth~~

seventh

eighth

ninth

tenth

Draw a picture.

Who is in seventh place?	Who is in ninth place?

Draw the animals in the squares to find the answers.

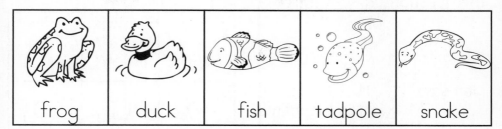

| frog | duck | fish | tadpole | snake |

1. The duck is first. The frog is between the tadpole and the duck. The fish is next to the tadpole. Where is the snake? _____

2. The frog is first. The tadpole is between the frog and the snake. The fish is fourth. Where is the duck? _____

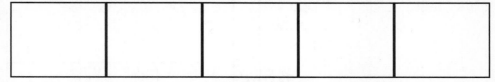

3. The duck is last. The tadpole is between the duck and the snake. The frog is second. Where is the fish? _____

4. The frog is second. The snake is last. The fish is between the duck and the snake. Where is the tadpole? _____

Word Problems Grade 2—RB–904012

Use the tally marks to fill in the mammal table. Then answer the questions.

	🦛	🐘	🐯
Jeff saw — II hippos, IIII elephants, III tigers	2	4	3
Brooke saw — I hippo, II elephants, III tigers			
Abby saw — III hippos, IIII elephants, I tiger			
Matt saw — I hippo, III elephants, II tigers			
TOTAL			

Jeff saw
II   hippos
IIII elephants
III   tigers

Brooke saw
I   hippo
II   elephants
III   tigers

Abby saw
III   hippos
IIII elephants
I   tiger

Matt saw
I   hippo
III   elephants
II   tigers

1. Who saw the most hippos?_____

2. Who saw the least elephants? _____

3. How many tigers were seen in all?_____

4. How many more elephants were seen than hippos? _____

Dolphin Feeding Schedule

Use the table to answer the questions. The dolphins eat half their fish in the morning and half at night.

Dolphin Feeding Schedule

Feeding Times: 8:00 a.m. and 5:00 p.m.

Dolphin	Amount of Food
Eddy	
Emma	
Eve	
Edmund	

Note: = 12 fish. = 6 fish

1. How many dolphins are at the zoo? _____

2. What times are the dolphins fed? _____

3. How many fish does Eddy eat in one day? _____

4. How many fish does Emma eat in the morning? _____

5. Which dolphin eats the most fish each day? _____

Word Problems Grade 2—RB–904012

Use the number sentences to solve the problem. Draw in the animals.

Picture Bank

hippo   monkey   kangaroo   lion

1. There were 10 hippos at the zoo. 5 more were born.
How many hippos are there?

10 + 5 = _____hippos

2. There were 9 monkeys at the zoo. 7 more came.
How many monkeys are there at the zoo?

9 + 7 = _____monkeys

3. There are 8 mother kangaroos at the zoo. Each has a baby in its pouch.
How many kangaroos are there altogether?

8 + 8 = _____kangaroos

4. There were 4 lions at the zoo. 7 more were born.
How many lions are at the zoo now?

4 + 7 = _____lions

10

What Bugs Me?

Read each problem. Draw a picture. Write a number sentence with the answer.

Picture Bank

spider   ladybug   fly   snail   bee

1. 5 bees flew to the hive from the roses. 9 more bees came from the dandelions.
How many bees were there altogether?

5 + 9 = 14 ___ bees

2. 4 snails crawled in the grass. 8 more came out of the weeds.
How many were there in all?

_____ snails

3. 6 flies fell into the spider's web. Right away 9 more fell into the web.
How many flies fell into the web?

_____ flies

4. 2 ladybugs were on a leaf. 6 more landed on the leaf. Then 7 more came.
How many ladybugs were on the leaf?

_____ ladybugs

   Word Problems Grade 2—RB–904012

Read each problem. Draw a picture to help you solve the problems.

1. There are 7 adult monkeys and 5 young monkeys in the cage. How many monkeys are in the cage?

2. Three monkeys in the cage are going to have babies. How many monkeys will there be when all the babies are born?

3. The largest monkey ate 9 bananas. The next largest monkey ate 8 bananas. How many bananas did they eat altogether?

4. There are 3 monkeys swinging in the tree. There are 6 monkeys swinging on the ropes. How many monkeys are swinging?

At the Reptile Cage

Read each problem. Draw a picture to help you solve the problem.

Picture Bank

lizard   snake   crocodile   turtle

1. One cage holds 6 types of lizards. Another cage holds 8 types. How many types of lizards are at the zoo?

_____

2. There are 4 turtles and 9 turtle eggs. If all the eggs hatch, how many turtles will there be?

_____

3. One snake is 9 inches long. Another snake is 7 inches long. If they line up in a straight line, how long would the line be?

_____

4. The crocodile has only 8 teeth on the top and 13 teeth on the bottom. How many teeth does he have in all?

_____

© Rainbow Bridge Publishing Word Problems Grade 2—RB–904012

Use the information above to solve the problems.

lizard	frog	turtle	monkey
Eats 2 flies each day.	Eats 3 ladybugs each day.	Eats 1 fish each day.	Eats 6 bananas each day.

1. Three lizards each eat flies. How many flies will they eat?

$$2 + 2 + 2 = 6$$

2. A frog eats ladybugs for 2 days. How many ladybugs will the frog eat?

_____

3. Six turtles are eating fish today. How many fish will they eat in all?

_____

4. Two monkeys ate bananas today. How many bananas did they eat?

_____

More Feeding the Animals

Use the chart to solve the problems. Write a number sentence with the answer.

Bales of Hay per Day

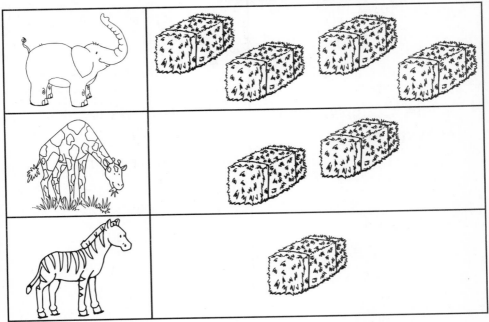

1. There are 4 elephants at the zoo. How many bales of hay do the elephants eat altogether? _____

2. There are 5 giraffes in the zoo. How many bales of hay do the giraffes eat? _____

3. How many bales of hay will the zebra eat each week? _____

4. How many bales of hay will 1 elephant, 1 giraffe, and a zebra eat in 1 day? _____

Word Problems Grade 2—RB–904012

Read each problem. Draw a picture to help you solve the problem.

There is one ostrich and one toucan. How many legs?

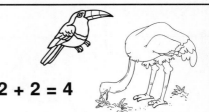

2 + 2 = 4

---

1. There are three mice. How many legs?

_____

2. There are two elephants and one mouse. How many legs?

_____

3. There is one ostrich, one spider and one elephant. How many legs?

_____

4. There are two toucans and two spiders. Which group has the most legs?

_____

Read each problem. Draw a picture to help you solve the problem.

Picture Bank

elephant   ostrich   toucan   tiger

1. There are 2 elephants and 3 ostriches in the parade. How many legs are there?

_____

2. One tiger has 9 stripes on his back. The other has 10 stripes. How many stripes are there altogether?

_____

3. There are 2 elephants, 3 kangaroos, 1 toucan, 2 tigers, and 3 lions in the parade. How many animals are in the parade?

_____

4. How many legs do 2 tigers and a toucan have altogether?

_____

17

Solve each number sentence. Draw a picture to help you.

Picture Bank

beetle   cocoon   butterfly   dragonfly   fly

---

1. The deer flipped his tail at 14 flies. Six of them flew away. How many are left?

$14 - 6 =$ _____ deer flies

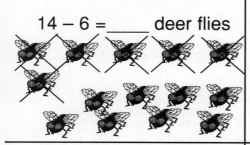

2. You find 13 beetles on a flower. A bird eats 7 of them. How many are left?

$13 - 7 =$ _____ beetles

---

3. While camping, 15 dragonflies land on your fishing pole. You brush 9 dragonflies off. How many are still on your pole?

$15 - 9 =$ __ dragonflies

4. There were 17 cocoons. Ten of them turned into butterflies. How many are still cocoons?

$17 - 10 =$ ___ cocoons

Solve each problem. Draw a picture to help you.

Picture Bank

pig   hen   cow   rabbit

---

1. A mother pig had 13 piglets. Seven piglets were taken to a farm. How many piglets were left in the zoo?

_____

2. A hen laid 11 eggs. Five eggs have hatched. How many eggs have not hatched?

_____

3. The cow had 15 flies on her back. She swatted away 7 of them with her tail. How many flies are still on her back?

_____

4. The rabbit had 12 babies. Only 4 babies were awake. How many baby rabbits were sleeping?

_____

Word Problems Grade 2—RB–904012

Bears and Balls

Use the picture to help you solve the problems. Write a number sentence with the answer.

There are 18 bears at the zoo.

1. There are 5 polar bears in the zoo. How many bears are not polar bears?

_____

2. The bears love to play with balls. Eight bears are playing with balls. How many are not playing with balls?

_____

3. The bears have 8 balls. They are red, blue, and yellow. Two balls are red, and four are blue. How many balls are yellow?

_____

4. There are 3 black bears and 4 brown bears in the zoo. How many bears are not black or brown?

_____

Solve the problems. Draw a picture to help you.

1. The zookeeper gave the apes 19 bananas. The apes ate 7. How many bananas are left?

_____

2. The biggest ape will eat 5 bananas. How many bananas will be left?

_____

3. There are 11 apes in the cage. Six apes ate bananas. How many apes did not eat?

_____

4. There are 9 female apes in the cage. How many of the 11 apes are male?

21

Birds of a Feather

Read each problem. Circle the + or – to tell whether you should add or subtract to solve the problem. Write a number sentence with the answer.

1. There are 7 canaries and 8 finches in the aviary. How many canaries and finches are there altogether?

+ –

_____

2. There are 9 birds flying and 6 birds sitting on a tree. How many birds are there altogether?

+ –

_____

3. There are 16 birds in one tree. Nine birds are red. How many birds are not red?

+ –

_____

4. There are 14 birds eating seeds. Eight birds flew away. How many birds are still eating seeds?

+ –

_____

Write the number sentence on the line.

The tarantula caught 11 grasshoppers for breakfast and 6 flies for lunch. How many munchies did he catch that day?

$$11 + 6 = 17$$

1. The tarantula has 7 potato beetles in his burrow. 7 more walked in. How many munchies does he have?

_____

2. There were 18 flies flying past the burrow. 5 flies got caught. How many flies did not get caught?

_____

3. There are 14 dragon-flies in the burrow with the tarantula. How many more dragonflies are there than tarantulas?

_____

4. If the tarantula catches 3 silverfish a day for 4 days, how many silverfish will he have altogether?

_____

Word Problems Grade 2—RB–904012

Creepy Crawlers

Write a number sentence to solve each problem. Think about whether you should add or subtract.

1. The ants were making a city. Seven ants were carrying crumbs, and 8 were carrying small sticks. How many ants were carrying something?

_____

2. A spider laid 15 eggs. Six have not hatched yet. How many eggs have hatched?

_____

3. The zoo has 12 centipedes and 7 scorpions. How many more centipedes are there than scorpions?

_____

4. There are 4 beetles, 6 ladybugs, and 3 spiders on one plant. How many bugs are on the plant?

_____

24

Penguins at Play

Solve each problem. The first problem is worked for you.

1. The zoo has 17 penguins. Nine penguins
are swimming. How
many penguins are not
swimming?

17 − 9 = 8

2. Seven penguins were
sliding on the ice.
How many of the
17 penguins were not
sliding?

_____

3. We saw 3 penguins
diving for fish,
2 penguins sliding, and
6 penguins walking.
How many penguins
did we see?

4. There are 8 penguins
standing by the rock
and 5 penguins
standing by the log.
How many penguins
are standing?

_____

25

Hens, Roosters, and Chicks

Write a number sentence on the line to solve each problem.

Picture Bank

hen   rooster   chick

1. The zoo has 23 chickens. Ten chickens are hens, and 10 are baby chicks. How many chickens are roosters?

   _____

2. Three of the zoo's 10 hens are white. The rest of the hens are brown. How many brown hens are there?

   _____

3. One hen has 9 eggs that have hatched. Four more eggs will hatch soon. How many chicks will the hen have when all her eggs have hatched?

   _____

4. One hen has 8 chicks. Another hen has 7 chicks. How many chicks do they have altogether?

   _____

Bird Watching

Follow the tree limbs to find which birds you will see. Write the birds you visit on the way.

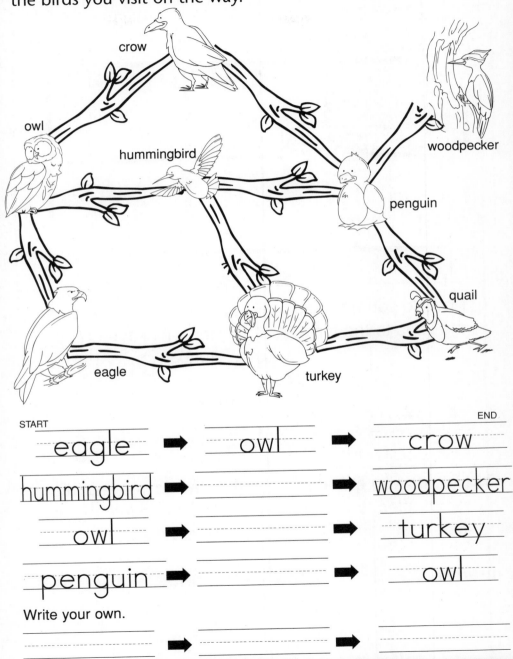

START				END
eagle	➡	owl	➡	crow
hummingbird	➡		➡	woodpecker
owl	➡		➡	turkey
penguin	➡		➡	owl

Write your own.

_____ ➡ _____ ➡ _____

Word Problems Grade 2—RB–904012

Zoo Visit

Use arrows to mark the direction you must go to get from place to place in the zoo.

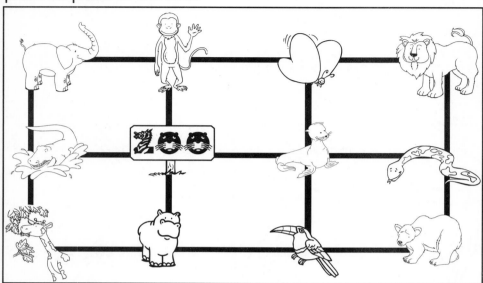

---

1. How do you get from the lions to the hippos?

lions   **hippos**

↓   ←   ←   ↓

2. How do you get from the monkeys to the toucan?

monkeys   **toucan**

__ __ __

---

3. How do you get from the giraffes to the lions?

giraffes   **lions**

__ __ __ __

4. How do you get from the elephants to the bears?

elephants   **bears**

__ __ __

What Temperature Is Best?

Write the temperature. Write warm or cold.

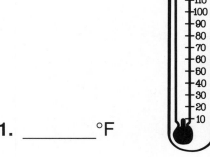

__40___ °F

___cold___

1. _____°F

_____

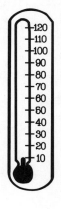

2. _____°F

_____

3. _____°F

_____

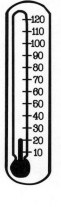

4. _____°F

_____

Word Problems Grade 2—RB–904012

What Temperature Is Best?

Circle the temperature that is the better estimate for the animals.

Lizard on a rock

(90°F)   50°F

1. Camel in the desert

100°F   30°F

2. Squirrels gathering nuts

89°F   44°F

3. Frogs on a pond

22°F   60°F

4. Penguins on an iceberg

20°F   52°F

© Rainbow Bridge Publishing

Read the sentences and complete the graph. Use the line graph to help you solve the following questions.

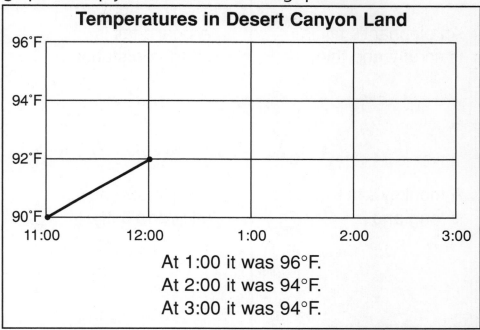

Temperatures in Desert Canyon Land

At 1:00 it was 96°F.
At 2:00 it was 94°F.
At 3:00 it was 94°F.

1. What was the temperature at 12:00?

_____

2. What time was the hottest?

_____

3. Was it warmer at 11:00 or 2:00?

_____

4. Was it cooler at 12:00 or 3:00?

_____

© Rainbow Bridge Publishing   Word Problems Grade 2—RB–904012

Read the sentences and answer the questions.

An elephant's tail is is scruffy and thin.

A bear's tail is short and stubby.

A monkey's tail is long and curly.

A squirrel's tail is thick and fluffy.

1. Who has the longest tail?

_____

2. Who has the shortest tail?

_____

3. Who has the thickest tail?

_____

4. Who has the thinnest tail?

_____

Answer the questions.

1. Which baby weighs the least?

_____

2. Which baby weighs the most?

_____

3. How many more pounds does the baby elephant weigh than the baby hippo?

_____

4. Which baby weighs more than any two other babies added together?

_____

Word Problems Grade 2—RB–904012

Look at the graph and complete the information below.

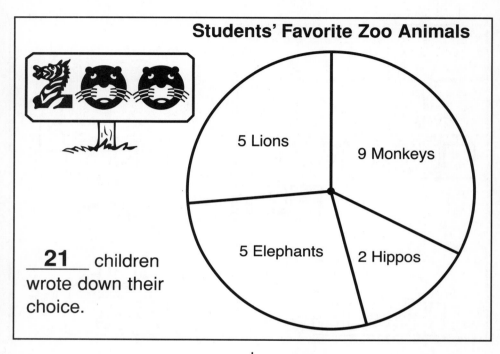

Students' Favorite Zoo Animals

5 Lions

9 Monkeys

5 Elephants

2 Hippos

___21___ children wrote down their choice.

1. _____ children like hippos.

2. More children like _____ than elephants.

3. Fewer children like _____ than lions.

4. The same number of children like elephants as _____.

Make your own circle graph of the animals students in your class like. Write your own questions.

Animals				
Tally marks				

Students' Favorite Zoo Animals

1.

2.

Word Problems Grade 2—RB–904012

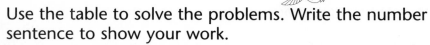

Use the table to solve the problems. Write the number sentence to show your work.

How much faster is the eagle than the duck?

100 − 70 = 30

Birds	Speed per Hour
hummingbird	120
eagle	100
hawk	90
duck	70
pigeon	60
seagull	50

1. Which bird flies the fastest?

   _____

2. Which bird flies the slowest?

   _____

3. If the seagull's speed was doubled, how fast would it be flying?

   _____

4. What is the difference between the hummingbird's speed and the pigeon's speed?

   _____

www.summerbridgeactivities.com

For the Birds

Answer the questions.

Bird Food

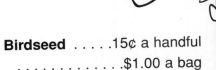

Sunflower Seeds . . .25¢ a handful
. . . .$1.25 a bag

Mixed Seed . . . . . . .10¢ a handful   **Birdseed** . . . . .15¢ a handful
. . . . .75¢ a bag   . . . .$1.00 a bag

---

1. How much would two bags of birdseed cost?

_____

2. If you had 75¢, how many handfuls of sunflower seeds could you buy?

_____

3. If you gave the clerk a quarter for one handful of birdseed, how much change would you get?

_____

4. If you gave the clerk a dollar bill for one bag of mixed seeds, how much change would you get?

_____

37

Answer the questions.

MENU

Sandwiches

Cheese	$1.00	Chicken	$1.75
Ham	$1.25	Hamburger	$1.50

Drinks

Milk	.50¢	Juice	.60¢
Soda	.40¢	Milk Shake	.75¢

Money Bank

1. You have $2.00. What will you order? How much change will you have left? Show us below.

2. Your friend has $3.00. What will she order? How much change will she have left? Show us below.

3. What is the difference between the cost of your lunch and your friend's? Write a number sentence.

4. What is the total cost of both meals? Write a number sentence.

We All Scream for Ice Cream!

Answer the questions.

Ice Cream

Flavors: Chocolate, Strawberry, Vanilla

1 scoop . . . . .50¢	2 scoops . . . . .75¢

add 10¢ for a waffle cone

fruit popsicle . . . . .50¢	fudgepop . . . . .75¢
ice cream bar $1.00	ice cream sandwich $1.00

1. How much would it cost to buy 1 scoop of ice cream for yourself and 3 friends?

2. If you paid for an ice cream bar and an ice cream sandwich with a five-dollar bill, how much change would you get?

3. How much would 2 scoops of vanilla ice cream on a waffle cone cost?

4. If you had $1.00, what would you buy, and how much change would you have left?

39

Buying Mammal Stickers

Answer the questions.

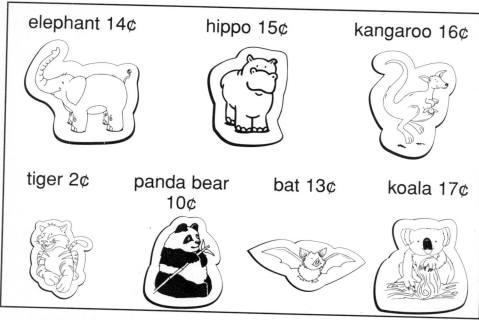

elephant 14¢　　hippo 15¢　　kangaroo 16¢

tiger 2¢　　panda bear 10¢　　bat 13¢　　koala 17¢

1. You have $1.00 to spend on stickers. Fill out the order form to show the stickers you want to buy.

Name of Sticker	How Many	Cost per One	Total Cost
Total cost for all the stickers			

2. Which sticker costs the most? _____

3. Which sticker costs the least? _____

4. Which sticker is your favorite? _____

Design your own shirt. Then complete the chart to figure out the cost of your shirt.

Shirt Types

Short Sleeve T-shirt $5.00

Long Sleeve T-shirt $7.00

Long Sleeve Sweatshirt . .$10.00

Long Sleeve
Hooded Sweatshirt $15.00

Extras

Zoo Logo Patch $2.00

Letters . . . . .10¢ per letter

Animal
Patches . . . . . .$1.00 per animal

Type of Shirt	Extras	Cost	Total

Total cost for all the shirts

Tiger Tales

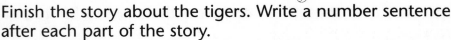

Finish the story about the tigers. Write a number sentence after each part of the story.

1. 10 tigers lay in the grass. 3 went to the water to get a drink. Then there were _____**7**_____ tigers left in the grass.

_____**10 – 3 = 7**_____

2. 5 more tigers came in from hunting. Now there are _____ tigers in the grass.

_____

3. 2 tigers left to find shade under a tree. How many tigers are still in the grass?

_____

4. 4 tigers came back from sleeping. Finally there are _____ tigers in the grass.

_____

www.summerbridgeactivities.com

© Rainbow Bridge Publishing

Elephant Trunk Tales

Finish the story about the elephants. Write a number sentence after each part of the story.

1. Once there were 12 elephants. Four of them packed their trunks to take a trip. _____ stayed home.

_____

2. The 4 elephants taking a trip each brought 2 hats. Together the elephants had _____ hats.

_____

3. The elephants taking the trip also packed 1 toothbrush each. Together the elephants had _____ toothbrushes.

_____

4. The elephants that stayed home each ate 2 peanuts. Together they finished a bag of _____ peanuts.

_____

43

Spots and Stripes Forever

Read the problem. Circle the + or − to tell whether you should add or subtract. Then write a number sentence with the answer.

1. The goat had 5 spots. The cow had 11 spots. How many more spots did the cow have?

+   −

_____

2. The zebra has 9 black stripes and 8 white stripes. How many stripes does it have altogether?

+   −

_____

3. The tiger has 19 stripes. Nine of the tiger's stripes are black. How many of its stripes are orange?

+   −

4. The hyena has 7 brown spots and 7 black spots. How many spots does it have altogether?

+   −

_____

All Aboard!

Read the problem. Circle the + or – to tell whether you should add or subtract. Then write a number sentence with the answer.

1. Eight people were on the zoo train. Four more people got on the train. How many were on the train?

+ –

_____

2. Seventeen people were on the train. Nine were children. How many adults were on the train?

+ –

_____

3. Seven adults and five children rode the train. How many people were on the train?

+ –

4. Seventeen people got on the train. Six people got off at the first stop. How many people stayed on the train?

+ –

_____

45

Count by twos to solve the problems. Draw a picture to show your thinking.

1. Fifteen birds sat on the branch. How many feet were on the tree?

_____

2. Thirteen people were riding the zoo bus. How many shoes were on the bus?

_____

3. Twenty-four rabbits hopped around the cage. How many ears can you count?

_____

4. Seventeen owls peeked out from the trees. How many eyes do you see?

_____

www.summerbridgeactivities.com

Count by fives. You are following the kangaroo. Write the number that tells where you land.

Start at 5. Hop forward 2 times. Where will you end up? __**15**__

backward   5   10   15   20   25   30   35   40   45   50   forward

1. Start at 35. Jump forward 3 times. What number are you on now?

_____

2. Start at 35. Hop backward 1 time. What number did you stop on?

_____

3. Start at 20. Hop forward 4 times. Where will you land?

_____

4. Start at 40. Jump backward 4 times. Where do you land?

_____

     Word Problems Grade 2—RB–904012

Count by tens to solve the problems.

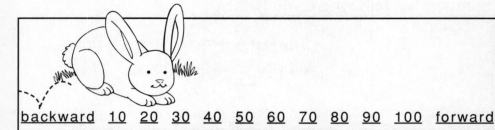

backward   10   20   30   40   50   60   70   80   90   100   forward

1. Start at 20. Hop forward 4 times. What number are you on now?

_____

2. Start at 80. Hop backward 2 times. Where did you land?

_____

3. Start at 40. Hop forward 3 times. Where did you land?

_____

4. Start at 10. Hop forward 8 times. What number did you land on?

_____

Taking Inventory

The vet was taking inventory of the supplies in the supply cabinet. Count by twos, fives, or tens to help complete the chart.

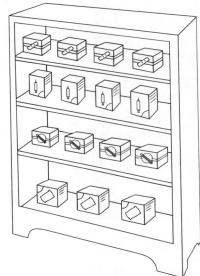

bandage wraps . . . . . .2 per box . . . . .stack of 8 boxes	thermometers . . . . . . .5 per box . . . . .stack of 10 boxes
syringes . . . . .10 per box . . . . .stack of 7 boxes	heating pads . . . . .2 per box . . . . .stack of 4 boxes

Supplies	Number per Box	Number of Boxes	Total
bandage wraps			
syringes			
thermometers			
heating pads			

Word Problems Grade 2—RB–904012

Look at the times for the zoo shows. Answer the questions.

Elephant Ride 1 hour 9:00 10:00 11:00 12:00	Tropical Gardens 15 minutes 9:15 10:15 	World of Flight Bird Arena 1 hour 9:00 10:00 11:00 12:00
Water Show 30 minutes 11:00	Petting Zoo 15 minutes 10:30 11:30 	Zoofari Express Train 45 minutes 9:30 10:30 11:30 12:30

1. If you go on an elephant ride at 11:00, what time will you be done?

_____

2. If you go on the Zoofari Train at 12:30, what time will you be done?

_____

3. What time will you be done at the Tropical Gardens if you go to the 10:15 show?

_____

4. Choose 3 shows. Make a schedule of when you will see each show and in what order.

_____

_____

_____

Shows at the Zoo

Look at the times for the zoo shows. Circle your answers.

Elephant Ride 1 hour 9:00 10:00 11:00 12:00	Tropical Gardens 15 minutes 9:15 10:15	World of Flight Bird Arena 1 hour 9:00 10:00 11:00 12:00

Water Show 30 minutes 11:00	Petting Zoo 15 minutes 10:30 11:30	Zoofari Express Train 45 minutes 9:30 10:30 11:30 12:30

1. If you go on the Elephant Ride at 9:00, will you be able to go to the Tropical Garden right after and make the 10:15 show?

Yes   No

2. If you go on the Zoofari Express Train at 10:30, will you be able to make it to the Water Show at 11:00?

Yes   No

3. If you go to the Petting Zoo at 11:30, will you be able to get to the Bird Arena to see the show by 12:00?

Yes   No

4. Find a classmate. Write down the time and name of an activity you would like to do together.

_____
_____
_____

Word Problems Grade 2—RB–904012

Use the watch to help you answer the questions.

1. The zoo opens at 10:00 a.m. and closes at 8:00 p.m. How many hours is the zoo open?

_____

2. If you get to the zoo at 10:00 and want to meet your friends for lunch at 12:30, how many hours will you have to look at animals?

_____

3. The Bird Show starts at 11:15. It lasts for 25 minutes. What time does it end?

_____

4. The Water Show starts at 11:00. It lasts for half an hour. What time does it end?

_____

You have 1 hour. Which animals will you be able to see?
Write the number sentence.

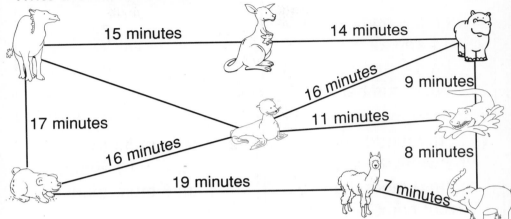

15 minutes

14 minutes

16 minutes

9 minutes

17 minutes

11 minutes

16 minutes

8 minutes

19 minutes

7 minutes

1. How many minutes will it take you to get from the camels to the llamas? _____ **17 + 19 = 36** _____

2. From the hippos, will it take longer to get to the bears or camels? _____Show why.

   _____

3. If you start at the seals and then visit the hippos and kangaroos, how many minutes will it take you?

   _____

4. Use your hour to see the zoo. Show which animals you can see in 60 minutes.

   _____ to _____ takes _____ minutes

   _____ to _____ takes _____ minutes

   _____ to _____ takes _____ minutes

   Total minutes _____

Word Problems Grade 2—RB–904012

Read and answer the questions.

I am a number between 14 and 22.
I am 4 more than 16.
What number am I? __**20**__

1. I am an odd number between 20 and 30.
I am 9 more than 12.
What number am I?

_____

2. I am a number between 10 and 20.
I am 6 more than 14 − 6.
What number am I?

_____

3. I am an odd number between 7 and 13.
I am 15 less than 26.
What number am I?

_____

4. I am an odd number between 9 and 15.
I am not the sum of 6 + 5.
What number am I?

_____

Koala Facts

Write a number sentence to solve each problem. Think about whether you should add or subtract the numbers.

1. Koalas eat 3 pounds of leaves each day. How many pounds of leaves do they eat each week?

_____

2. Adult koalas weigh about 22 pounds. A young koala weighs 7 pounds. How many pounds will the young koala probably gain?

_____

3. Five koalas are eating leaves. Six are sleeping. Three are climbing. How many koalas are there in all?

4. Koalas sleep about 16 hours each day. In a 24-hour day, how many hours is a koala awake?

Word Problems Grade 2—RB–904012

Solve each problem. Think about whether you should add or subtract. Draw a picture to help you think.

1. Nineteen bats were hanging. Fifteen flew away. How many were left?

2. Bats have 1 baby per year. How many years will it take for 10 bats to have 30 babies?

3. Brown bats can live to be 32 years old. If a brown bat is 20 years old, how many more years might it live?

4. A brown bat can eat 600 mosquitoes in an hour. How many mosquitoes can it eat in 2 hours?

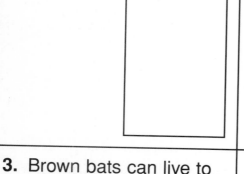

Answer the questions.

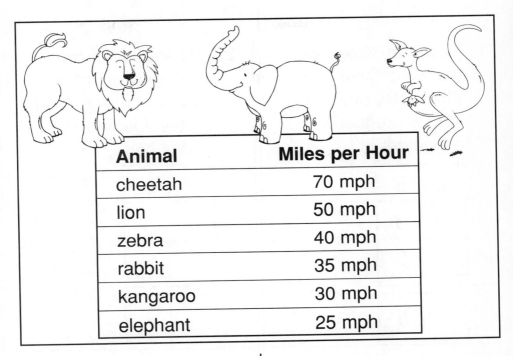

Animal	Miles per Hour
cheetah	70 mph
lion	50 mph
zebra	40 mph
rabbit	35 mph
kangaroo	30 mph
elephant	25 mph

1. How much faster does a cheetah run than an elephant?

2. How much faster does a lion run than a rabbit?

3. How many miles could a lion run in 2 hours?

4. How many hours would it take an elephant to run 100 miles?

57

Stripes and Spots

Read the questions. Solve the problems.

1. There are 3 zebras that each have 25 stripes. How many stripes are there altogether?

$$
\begin{array}{r}
2\;5 \\
2\;5 \\
+\;2\;5 \\
\hline
7\;5
\end{array}
$$

2. 3 tigers each have 22 stripes on them. How many stripes are on the tigers altogether?

$$
\begin{array}{r}
\; \\
\; \\
+\;\; \\
\hline
\;
\end{array}
$$

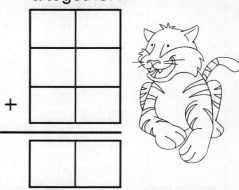

3. If there are 2 snakes that have 15 white stripes, how many stripes do they have altogether?

$$
\begin{array}{r}
\; \\
+\;\; \\
\hline
\;
\end{array}
$$

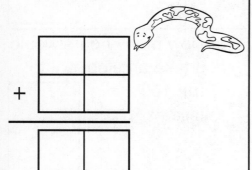

4. A raccoon's tail has 4 black and 4 white stripes. If there are 4 raccoons, how many black stripes do they have?

☐ + ☐ + ☐ + ☐ = ☐

Use the information in the table to help you solve the problems.

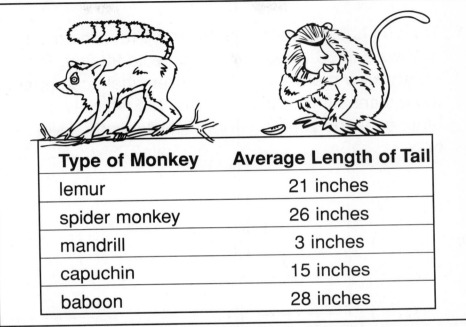

Type of Monkey	Average Length of Tail
lemur	21 inches
spider monkey	26 inches
mandrill	3 inches
capuchin	15 inches
baboon	28 inches

1. Which monkey has the longest tail?

2. How much longer is the baboon's tail than the capuchin's tail?

3. How much longer is the spider monkey's tail than the lemur's tail?

4. How much longer is the capuchin's tail than the mandrill's tail?

Word Problems Grade 2—RB–904012

Animal Math

Solve each problem. Think about whether you should add or subtract.

1. The zoo has 47 types of birds and 58 types of fish. How many more types of fish than birds does the zoo have?

2. The blue whale can grow to 98 feet. The killer whale can grow to 32 feet. How much smaller is the killer whale?

3. An albatross egg takes 84 days to hatch. A wren's egg hatches in 12 days. How much longer does it take an albatross to hatch than a wren?

4. The flying frog sailed 22 feet from one tree to another. Then it sailed 17 feet to another tree. How many feet did the frog fly?

Eagle Eyes

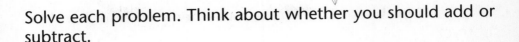

Solve each problem. Think about whether you should add or subtract.

1. There are 60 kinds of eagles in the world. The zoo has 10 different kinds of eagles. How many kinds of eagles do not live in the zoo?

2. The zoo fed the eagles 24 snakes on Monday and 32 snakes on Thursday. How many snakes did the eagles eat?

3. A golden eagle can grow to 39 inches. A bald eagle can grow to 31 inches. How much larger can the golden eagle grow than the bald eagle?

4. The bald eagle used 44 sticks and 25 twigs to build a nest. How many sticks and twigs were used altogether?

Word Problems Grade 2—RB–904012

Solve each problem. Think about whether you should add or subtract.

1. The spotted salamander had 23 spots on its back and 16 spots on its tail. How many spots did it have altogether?

2. A tiger salamander was 2 inches long. By the time it stopped growing it was 15 inches long. How many inches did it grow?

3. The salamander ate 27 earthworms and 12 ants. How many things did the salamander eat?

4. If 4 salamanders each laid 100 eggs, how many salamander eggs would there be?

Frog Findings

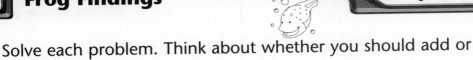

Solve each problem. Think about whether you should add or subtract.

1. The frog laid 48 eggs. A fish ate 23 of the eggs. How many eggs were left?

2. A frog jumped 17 inches in its first hop and 16 inches in its second hop. How far did the frog jump?

3. One frog had 38 spots on its back. Another frog had 41. How many spots did they have altogether?

4. One cage had 15 frogs. The other cage had 22 frogs. How many more frogs were in the second cage?

Word Problems Grade 2—RB–904012

Write a multiplication sentence to solve each problem.

How many eyes?

___**5**___ x ___**2**___ = ___**10**___

1. How many tails?

____ x ____ = ____

2. How many ears?

____ x ____ = ____

3. How many legs?

____ x ____ = ____

4. How many wings?

____ x ____ = ____

www.summerbridgeactivities.com © Rainbow Bridge Publishing

How Many?

Write a multiplication sentence to solve each problem.

1. How many humps?

_____ x _____ = _____

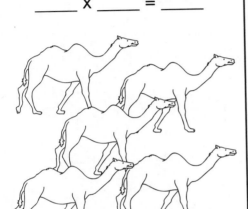

2. How many legs?

_____ x _____ = _____

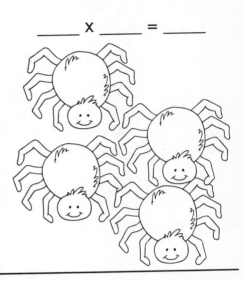

3. How many ears?

_____ x _____ = _____

4. How many whiskers?

_____ x _____ = _____

Word Problems Grade 2—RB–904012

Write a multiplication sentence to solve each problem.

1. Each rabbit stands for six babies. How many baby rabbits?

_____ x _____ = _____

2. Each lamb stands for two babies. How many lambs?

_____ x _____ = _____

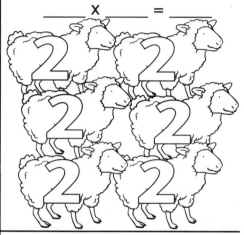

3. Each duck stands for four babies. How many baby ducks?

_____ x _____ = _____

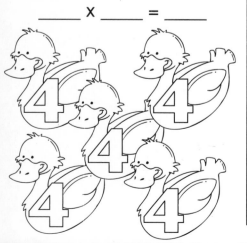

4. Each elephant stands for one baby. How many baby elephants?

_____ x _____ = _____

On Your Toes

Write a multiplication sentence to solve each problem.

1. Gorillas have 10 toes. How many toes do 8 gorillas have?

_____ X _____ = _____

2. Cats have 8 toes. How many toes do 4 cats have?

_____ X _____ = _____

3. Camels have 4 toes. How many toes do 5 camels have?

_____ X _____ = _____

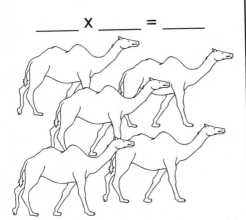

4. Bears have 10 toes. How many toes do 6 bears have?

_____ X _____ = _____

Word Problems Grade 2—RB–904012

Animal Count

The zookeeper is finishing counting all the animals. Help him figure out the number of each animal in the zoo, and complete the table below.

There are 8 tigers.

There are 4 more lions than tigers.

There are 3 fewer elephants than lions.

There are 6 fewer alligators than elephants.

There are 5 more monkeys than lions.

There are 9 more birds than monkeys.

Animal	How Many?
tigers	
lions	
elephants	
alligators	
monkeys	
birds	

Field Trip Favorites

After their field trip to the zoo, Mrs. Flannery's class voted on their favorite animal. Figure out how many students liked each type of animal best. Complete the table.

There are 27 students in the class.

8 students liked the monkeys best.

5 fewer students liked giraffes than monkeys.

Twice as many students liked the tigers as liked the giraffes.

One less student liked lions than tigers.

Two more students liked penguins than giraffes.

Animal	Number of Students
monkeys	
giraffes	
tigers	
lions	
penguins	

Word Problems Grade 2—RB–904012

Zoo News

Read the news articles carefully, and decide which information is needed to solve the problem. Cross out any information that is not necessary. Then solve the problem.

1. Yesterday, 10 crocodiles moved into a new cage. Today, 13 more crocodiles moved into the cage. The biggest was 13 feet long. How many crocodiles are in the new cage?

_____

2. A baby gorilla was born on October 6. It weighed 22 pounds. Its mother weighed 327 pounds. It was born at 6:00 a.m. How much more does the mother weigh than her baby?

_____

3. Twenty-two pigeons flew the coop after a zookeeper left a cage open. The zookeeper has worked at the zoo 12 years. Ten pigeons have returned. How many pigeons are still loose?

_____

4. 57 more people visited the zoo on Thursday than on Friday. 32 visitors were children. 511 people visited the zoo on Friday. How many people visited the zoo on Thursday?

_____

More Zoo News

Read the news articles carefully, and decide which information is needed to solve the problem. Cross out any information that is not necessary. Then solve the problem.

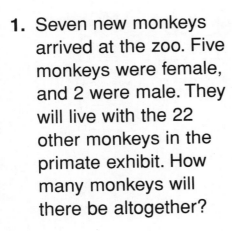

1. Seven new monkeys arrived at the zoo. Five monkeys were female, and 2 were male. They will live with the 22 other monkeys in the primate exhibit. How many monkeys will there be altogether?

   _____

2. Jen Jones is the new dolphin trainer. She is 39 years old. She worked at the Brooklyn Zoo for 12 years and the San Diego Zoo for 7 years. How many years of zoo experience does Jen have?

   _____

3. Three mother sheep gave birth to twins on Tuesday. Three other sheep are expecting babies soon. How many baby lambs were born on Tuesday?

   _____

4. The polar bear is 28 years old. The polar bear came to the zoo 22 years ago. Seven other bears live at the zoo. How old was the polar bear when it came to the zoo?

   _____

Critical Thinking Skills

Word Problems Grade 2—RB–904012

Animal Reports

Andy, Beth, Lisa, and Gary are doing reports on zoo animals. But whose report is whose? Read the clues.

Andy's animal has stripes.

Lisa's animal is black and white.

Beth's animal does not have a mane.

Put an X in the box when you know an animal does not belong in the report. Put an O when you know the animal does belong.

	Tiger	Zebra	Lion	Giraffe
Andy				
Beth				
Lisa				
Gary				

Animal Mix Up

The zookeeper left the doors to 4 cages open, and the animals got out. Help the zookeeper put the animals back into their cages.

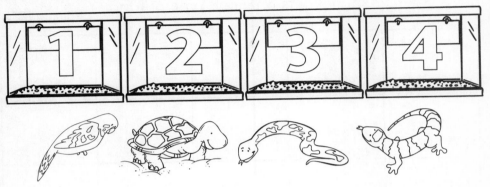

The animal that belongs in cage number 1 has feathers.

Neither the snake nor the lizard belongs in cage number 2.

The snake is between the turtle and the lizard.

Put an X in the box when you know an animal does not belong in the cage. Put an O when you know the animal does belong.

	1	2	3	4
snake				
bird				
turtle				
lizard				

73

Feeding Frenzy

Put an X in the box when you know an animal does not get that food. Put an O when you know the animal does get that food. Each animal gets only one type of food.

Zebras do not eat leaves.

A giraffe eats more than a koala.

Elephants eat only large amounts of hay.

Answer Pages

Page 3
1. Thursday
2. picture of heart
3. 1
4. 11
5. 2

Page 4
1. Wednesday
2. Bird Show
3. No
4. 2:00

Page 5

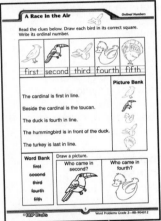

Page 6

Page 7

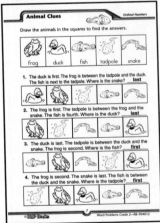

Page 8
1. Abby
2. Brooke
3. nine
4. six

Word Problems Grade 2—RB–904012

Answer Pages

Page 9
1. 4
2. 8:00 a.m. and 5:00 p.m.
3. 24 fish
4. 15
5. Edmund

Page 10
1. 15 hippos
2. 16 monkeys
3. 16 kangaroos
4. 11 lions

Page 11
1. 5 + 9 = 14 bees
2. 4 + 8 = 12 snails
3. 6 + 9 = 15 flies
4. 2 + 6 + 7 = 15 ladybugs

Page 12
1. 12 monkeys
2. 15 monkeys
3. 17 bananas
4. 9 monkeys

Page 13
1. 14 types of lizards
2. 13 turtles
3. 16 inches long
4. 21 teeth

Page 14
1. 2 + 2 + 2 = 6
2. 3 + 3 = 6
3. 1 + 1 + 1 + 1 + 1 + 1 = 6
4. 6 + 6 = 12

Page 15
1. 4 + 4 + 4 + 4 = 16
2. 2 + 2 + 2 + 2 + 2 = 10
3. 1 + 1 + 1 + 1 + 1 + 1 + 1 = 7
4. 4 + 2 + 1 = 7

Page 16
1. 4 + 4 + 4 = 12
2. 4 + 4 + 4 = 12
3. 2 + 8 + 4 = 14
4. 2 + 2 = 4, 8 + 8 = 16, spiders

Page 17
1. 4 + 4 + 2 + 2 + 2 = 14 legs
2. 9 + 10 = 19 stripes
3. 2 + 3 + 1 + 2 + 3 = 11 animals
4. 4 + 4 + 2 = 10 legs

Page 18
1. 8 deer flies
2. 6 beetles
3. 6 dragonflies
4. 7 cocoons

Page 19
1. 13 − 7 = 6 piglets
2. 11 − 5 = 6 more eggs
3. 15 − 7 = 8 flies
4. 12 − 4 = 8 rabbits

Page 20
1. 18 − 5 = 13 are not polar bears
2. 18 − 8 = 10 are not playing with balls
3. 8 − 6 = 2 balls are yellow
4. 18 − 7 = 11 bears are not black or brown

Page 21
1. 19 − 7 = 12 bananas
2. 19 − 5 = 14 bananas
3. 11 − 6 = 5 apes
4. 11 − 9 = 2 apes

Page 22
1. +, 7 + 8 = 15
2. +, 9 + 6 = 14
3. −, 16 − 9 = 7
4. −, 14 − 8 = 6

Page 23
1. 7 + 7 = 14
2. 18 − 5 = 13
3. 14 − 1 = 13
4. 3 + 3 + 3 + 3 = 12

Page 24
1. 7 + 8 = 15
2. 15 − 6 = 9
3. 12 − 7 = 5
4. 4 + 6 + 3 = 13

www.summerbridgeactivities.com © Rainbow Bridge Publishing

Answer Pages

Page 25
1. $17 - 9 = 8$
2. $17 - 7 = 10$
3. $3 + 2 + 6 = 11$
4. $8 + 5 = 13$

Page 26
1. $23 - 20 = 3$
2. $10 - 3 = 7$
3. $9 + 4 = 13$
4. $8 + 7 = 15$

Page 27 (Answers may vary.)
eagle, owl, crow
hummingbird, penguin, woodpecker
owl, eagle, turkey
penguin, hummingbird, owl

Page 28 (Answers may vary.)
1. down, left, left, down
2. right, down, down
3. right, right, right, up, up
4. down, down, right, right, right

Page 29
1. 10°F, cold
2. 90°F, warm
3. 20°F, cold
4. 70°F, warm

Page 30
1. 100°F
2. 44°F
3. 60°F
4. 20°F

Page 31

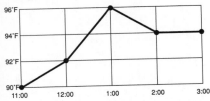

1. 92°F
2. 1:00
3. 2:00
4. 12:00

Page 32
1. monkey
2. bear
3. squirrel
4. elephant

Page 33
1. monkey
2. elephant
3. 20 pounds
4. elephant

Page 34
1. 2
2. monkeys
3. hippos
4. lions

Page 35
Answers will vary.

Page 36
1. hummingbird
2. seagull
3. 100
4. 60

Page 37
1. $2.00
2. 3 handfuls
3. 10¢
4. 25¢

Page 38
1–4. Answers will vary.

Page 39
1. $2.00
2. $3.00
3. 85¢
4. Answers will vary.

Page 40
1. Answers will vary.
2. koala
3. tiger
4. Answers will vary.

© Rainbow Bridge Publishing   Word Problems Grade 2—RB–904012

Answer Pages

Page 41
1–4. Answers will vary.

Page 42
1. 7, 10 − 3 = 7
2. 12, 7 + 5 = 12
3. 12 − 2 = 10
4. 14, 10 + 4 = 14

Page 43
1. 8, 12 − 4 = 8
2. 8, 2 + 2 + 2 + 2 = 8
3. 4, 1 + 1 + 1 + 1 = 4
4. 16, 2 + 2 + 2 + 2 + 2 + 2 + 2 + 2 = 16

Page 44
1. −, 11 − 5 = 6
2. +, 9 + 8 = 17
3. −, 19 − 9 = 10
4. +, 7 + 7 = 14

Page 45
1. +, 8 + 4 = 12
2. −, 17 − 9 = 8
3. +, 7 + 5 = 12
4. −, 17 − 6 = 11

Page 46
1. 30 feet
2. 26 shoes
3. 48 ears
4. 34 eyes

Page 47
1. 50
2. 30
3. 40
4. 20

Page 48
1. 60
2. 60
3. 70
4. 90

Page 49

Supplies	# per Box	# of Boxes	Total
bandage wraps	2	8	16
syringes	10	7	70
thermometers	5	10	50
heating pads	2	4	8

Page 50
1. 12:00
2. 1:15
3. 10:30
4. Answers will vary.

Page 51
1. yes
2. no
3. yes
4. Answers will vary.

Page 52
1. 10 hours
2. 2 1/2 hours
3. 11:40
4. 11:30

Page 53
1. 17 + 19 = 36
2. bears, 16 + 16 = 32 > 14 + 15 = 29
3. 16 + 14 = 30
4. Answers will vary.

Page 54
1. 21
2. 14
3. 11
4. 13

Page 55
1. 3 + 3 + 3 + 3 + 3 + 3 + 3 = 21
2. 22 − 7 = 15
3. 5 + 6 + 3 = 14
4. 24 − 16 = 8

Page 56
1. 19 − 15 = 4
2. 3 years
3. 32 − 20 = 12
4. 600 + 600 = 1,200

Answer Pages

Page 57
1. 45 mph
2. 15 mph
3. 100 miles
4. 4 hours

Page 58
1. 25 + 25 + 25 = 75
2. 22 + 22 + 22 = 66
3. 15 + 15 = 30
4. 4 + 4 + 4 + 4 = 16

Page 59
1. baboon
2. 13 inches
3. 5 inches
4. 12 inches

Page 60
1. 58 − 47 = 11
2. 98 − 32 = 66
3. 84 − 12 = 72
4. 22 + 17 = 39

Page 61
1. 60 − 10 = 50
2. 24 + 32 = 56
3. 39 − 31 = 8
4. 44 + 25 = 69

Page 62
1. 23 + 16 = 39
2. 15 − 2 = 13
3. 27 + 12 = 39
4. 100 + 100 + 100 + 100 = 400

Page 63
1. 48 − 23 = 25
2. 17 + 16 = 33
3. 38 + 41 = 79
4. 22 − 15 = 7

Page 64
1. 6 x 1 = 6
2. 5 x 2 = 10
3. 4 x 4 = 16
4. 5 x 2 = 10

Page 65
1. 5 x 2 = 10
2. 4 x 8 = 32
3. 3 x 2 = 6
4. 4 x 4 = 16

Page 66
1. 4 x 6 = 24
2. 6 x 2 = 12
3. 5 x 4 = 20
4. 3 x 1 = 3

Page 67
1. 8 x 10 = 80
2. 8 x 4 = 32
3. 4 x 5 = 20
4. 6 x 10 = 60

Page 68

Animal	How Many?
tigers	8
lions	12
elephants	9
alligators	3
monkeys	17
birds	26

Page 69

Animals	Number of Students
monkeys	8
giraffes	3
tigers	6
lions	5
penguins	5

© Rainbow Bridge Publishing   Word Problems Grade 2—RB–904012

Answer Pages

Page 70

1. ~~The biggest crocodile was 13 feet long.~~
   23 crocodiles
2. ~~It was born at 6:00 in the morning.~~
   305 pounds more
3. ~~The zookeeper worked at the zoo 12 years.~~
   12 pigeons
4. ~~32 of those people were children.~~
   568 people

Page 71

1. ~~Five monkeys were female, and 2 were male.~~
   29 monkeys
2. ~~She is 39 years old.~~
   19 years
3. ~~Three other sheep are expecting babies soon.~~
   6 lambs
4. ~~Seven other bears live at the zoo.~~
   6 years old

Page 72

	Tiger	Zebra	Lion	Giraffe
Andy	O	X	X	X
Beth	X	X	X	O
Lisa	X	O	X	X
Gary	X	X	O	X

Andy did a report on tigers.
Beth did a report on giraffes.
Lisa did a report on zebras.
Gary did a report on lions.

Page 73

	1	2	3	4
snake	X	X	O	X
bird	O	X	X	X
turtle	X	O	X	X
lizard	X	X	X	O

The bird belongs in cage 1.
The turtle belongs in cage 2.
The snake belongs in cage 3.
The lizard belongs in cage 4.

Page 74

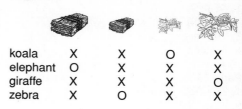

	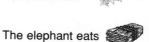			
koala	X	X	O	X
elephant	O	X	X	X
giraffe	X	X	X	O
zebra	X	O	X	X

The koala eats .

The elephant eats .

The giraffe eats .

The zebra eats .

www.summerbridgeactivities.com